THE COMPLETE

NFT

HANDBOOK

THE SIMPLE GUIDE TO CREATE, SELL AND BUY NFTS (NON-FUNGIBLE TOKENS)

BY

WILLIAM ROSS

ISBN: 978-1-63809-024-3

Disclaimer

Please note, the information contained in this book, are for educational purposes only. Every attempt has been made to provide accurate, up to date and reliable complete information. By reading this document, the reader agrees that

under no circumstances are we responsible for any losses, direct or indirect, which are incurred as a result of the use of the information contained in this document, including but not limited to errors, omissions or inaccuracies.

AF879993

TABLE OF CONTENTS

INTRODUCTION

Only a few situations are more ironic than this in the techno-economic and digital assets world. The more frequent talks and analyses of a concept, idea, practice, or situation, the clearer it becomes; that's what simple logic dictates and not the otherwise. However, I have now seen that this is not always the case.

I have volunteered to rise to this challenge and take you, regardless of the level of your exposure to NFTs, through the peripheral and nitty-gritty. I have seen how NFTs can be a life-changer for many. I realize that the impact I have been making while coaching individuals and groups about this can be greater. Many more could benefit from the knowledge of NFTs if I endeavor to reach a larger audience outside webinars and videoconferencing.

That's the inspiration behind the writing of this book on NFTs so that you could have the same taste of milk and honey these tokens are flowing in. Yes, this book will benefit beginners. Yet, the learned and experienced investors have a lot to learn in this material with the potential to hasten and enhance their profit.

Oh! Excuse me. What am I talking about? I probably need to start by reminding the beginners among my readers what NFTs as an acronym implies. Simply stated: NFTs means non-fungible tokens; yes tokens that aren't fungible, replaceable, or interchangeable. Okay? But I would love to hold it there for now.

This book tells you all you need to know about NFTs. I purposely package the content to appeal to everyone with each chapter building on the previous so that the knowledge base of NFTs can be piled up conveniently.

For instance, the first chapter introduces you to NFTs, covering aspects such as what exactly they are, why people collect them, types and aspects, and their extrinsic elements. In the second chapter, I dwell on the true value of NFTs dealing with aspects such as why collectibles have value, the problems with traditional art, the problems with collectibles and memorabilia, digital art before NFTs, their true advantages, and external forces that drive value. The two chapters that follow expose the NFTs history and marketplaces while chapters 5 and 7 deal with how to create, mint, sell, and buy NFTs. Their legal aspects and future dominate the discussion in the last two chapters.

It's my sincere hope that this book on the basics of NFTS will change your story and add your sweet experience to the ones already being told by gainers in NFTS.

CHAPTER 1: BASICS OF NFTS

I feel good beginning this chapter with the assumption that a beginner will want to know the basics of NFTS, starting from the acronym, which stands for non-fungible tokens.

What Exactly Are NFTs?

NFTs are basically cryptographic assets on a blockchain. The tokens have unique identification codes and metadata distinguishing each one from another. Please don't get carried away by the mention made of crypto and blockchain. No, they're not cryptocurrencies. They can neither be traded nor exchanged at equivalency. Another difference between NFTs and cryptocurrencies is reflected in their descriptive compound name, "non-fungible." This clearly differentiates them from fungible tokens and identical cryptocurrencies, which can comfortably serve as a medium for the exchange of goods and services in commercial transactions.

NFTs are one way of physically registering or representing precious items, like real estate and one-of-a-kind artwork, video, or any other form of digital or, of course, physical item, on a blockchain. Since NFTs are based on blockchains, they can also be an appropriate way to represent individuals' identities, property rights, and more. They're thus a good way to avoid intermediaries and connect artists with their audiences or for identity management.

Just an example of this situation is what took place in early March 2021 when a group of NFTs by Mike Winklemann, the renowned digital artist, better known as Beeple, sold for over 69 million US dollars. It was a crafted composite collage comprised of Beeple's First 5000 Days. That sale set a precedent and a record as the most expensive piece of digital art ever sold.

You can analyze NFTs by breaking the descriptive acronym into two, namely, non-fungible and token. A fungible item is known to be interchangeable. A non-fungible item is a unique item that can't be replaced or replicated. For instance, the original *Mona Lisa* painting in the physical or digital museum, that tweet of yours, the video you create, or other pieces of artwork are unique and are thus non-fungible.

Since NFTs are bought and sold online, the transaction is frequently with cryptocurrency having been generally encoded with the same underlying software. Of course, NFTs have been around since 2014, they're recently gathering momentum because of their increasing popularity. So, a staggering 174 million US dollars has been spent on NFTs since November 2017.

Why People Collect NFTs

The words of a blockchain expert and fintech professor at Rutgers Business School, Dr. Merav Ozair, are quite instructive in understanding why people collect NFTs: "The power of NFTs is the authentication. If you think about it, everything in our economy is transaction-based, even an email, a payment or sending money, these are all transactions and for them to be official, they need to be authenticated." For the reason given by this don, there'll always be the need for NFTs.

NFTs bypass third parties or intermediaries and have the authenticity of a work, payment, a license, and such things declared. This market efficiency is probably the most obvious benefit of NFTs. You can also deduce that this is a way out for content creators such as artists, musicians, and writers who hitherto must go through intermediaries to make sales.

Royalty issues have been effectively addressed. To put it in context; a successful *YouTuber* who sings or produces videos might make huge bucks each month from advertising revenues. However, YouTube itself always takes a significant cut that usually outweighs what the content creator makes. See what NFT tokens can help you avoid.

As a physical asset can easily be converted into a digital one, the processes have been streamlined. NFTs are also being used in the exchange of digital tokens that link to a digital file asset. While it can't be said that the practice of digitally representing physical assets is neither novel nor of unique identification in use, it can be a strong agent of change when combined with the benefits of a tamper-resistant blockchain of smart contracts.

People run after non-fungible tokens for their excellent identity management. This can be illustrated with the case of physical passports that must be produced at every entry and exit point. Once individual passports have been converted into NFTs, with their unique identifying characteristics, you've streamlined and simplified the entry and exit processes for jurisdictions.

Those seeking to democratize investing by fractionalizing their physical assets such as real estate, and artworks, want to collect NFTs since it makes it much easier to divide digital real estate assets among multiple owners. That tokenization ethic need not be constrained to real estate. Artwork and other videos too can have multiple owners while each owner has ownership of a digital equivalent of a fraction of the work. One big plus here is that the arrangement is capable of increasing the worth and revenue from the work.

An example of this is *Mass Banner* which is a unique proposal. Even by the complex world of NFTs, this NFT has been split into 266,445 shares which were bought up by some 30,000 buyers. Thus, no single individual owns the entire project with a total value of $91.8 million.

From what we've discussed up to this level, you'd see that NFTs are also a store of value. They can represent something of value such as a sporting moment or a landed property. Why do you think people are now buying the CryptoPunks NFT series which has been attracting in recent times individual sales in millions of dollars? It's because they believe that it'll be worth more in the future. To that extent, NFTs can also be considered an investment product since the price today is much lower than what the investor believes the NFT could be worth in the future.

A good example of successful investment in NFT tokens is the case of the $4.3 million purchase of 100 virtual islands that you've likely heard about in the *Sandbox Metaverse*. Granted, this might sound like an outrageous amount to invest in digital real estate. You know what? The buyer is already creating virtual villas on each island, and 90% of them are sold out within a day at $15,000 each. And don't have your mind blown up by this; many of these virtual villas have already been relisted at over $100,000.

Doesn't this tell you that NFTs that tokenize virtual plots of land are already being sought after to develop digital real estate for profit? In theory, digital realtors are here, moving the traditional real estate development industry into its own niche via the Metaverse. All these happen without traditional barriers such as cumbersome verification processes and high fees involved in trying to buy real estate in another country.

Another reason why many people collect NFTs is that there are small entry barriers. In this industry, investment spaces are open to those on a budget. While I have given examples of where people are paying millions of dollars to access a single NFT, some NFTs can be acquired with a few tens of thousands of dollars.

NFTs are now being used as collateral as some innovators can now allow you to access financing for pledging your NFT token as collateral. This bears some semblance with what obtains in conventional secured loans in which case the borrower will put up an asset of a certain amount of money to access it. What a borrower is putting up in the case of crypto loans is an NFT that has real-world value. The approval of the lending agreement is not only almost instantly but also requires no credit checks. If the borrower defaults, after all, the crypto lending site simply sells the token.

NFTs have a huge marketplace. While the estimates on the future total value of the NFT marketplace vary depending on the source, we can see the huge amount of fiat money that has changed hands in recent years. In 2021 alone, for example, Bloomberg reported the sale of almost $41 billion worth of NFTs. Optimism is high that the growth of the NFT space is not about to abate. Going by that, the industry has the potential to cross the mark of the $1-trillion market in no distant time.

Another example is *Yuga Labs*, the company credited with the creation of the *Bored Ape Yacht Club* NFT collection owned by Eminem, Jimmy Fallon and Paris Hilton. *Yahoo Finance* reports that following rumors of investment by Silicon Valley venture capital firm Andreessen Horowitz, the company has been valued at $5.

Types of NFTs

It's getting more interesting as you're now making a bit more sense of how NFTs work, right? How many different types of non-fungible tokens are currently on the market? I'll be giving you more examples of non-fungible tokens that have been sold in recent years.

- **Physical Real Estate**

This is perhaps the most fascinating type of NFT covering both physical and virtual properties. You hardly find two properties with the same features. This makes physical real estate a perfect example of an asset that can be tokenized. Recently, a 4-bedroom house in Florida was sold for over $653,000 as an NFT and was paid for with Ethereum. Since the house is being represented as an NFT token, if the owner wishes to sell it, it's a matter of a wallet-to-wallet transaction. It's already on the Ethereum blockchain.

- **Virtual Real Estate in the Metaverse**

The Metaverse is a digital version of the real world. Everything taking place in the material world is being represented in the Metaverse and this explains the expected exponential growth of interest in the space in the coming years. This should also give you an idea of the reason why the two terms NFTs and the Metaverse often collocate in discussions. An example of this is what plays out in Decentraland, a 3D gaming world allowing players to buy virtual land and then build real estate there. Each plot of land in this game is subsequently represented by a unique NFT token sellable on the open market.

Consider what happened in November 2021 when a plot of land in *Axie Infinity*, a Metaverse gaming, sold for over US$2.3 million. About a month later, a plot of 100 virtual islands on the *Sandbox* platform, another popular Metaverse game, sold for over US$4.3 million. All these show that real estate NFTs are becoming popular in the Metaverse.

- **CryptoPunks and Other Digital Collectables**

Let's talk about the CryptoPunks token which you're likely to have come across in your search. CryptoPunks is a representation of a collection of 10,000 digital characters created in 2007. There is an NFT token backing up each CryptoPunk which is like no other. One of the most expensive CryptoPunk NFTs out there, for now, was sold in February 2022 for some $23 million being the value of 8,000 Ethereum. It is *Larva Labs,* which was sold by its original owner in 2017 for only $1,646 (also corroborating the fact that NFTs are a store of value). Thousands of other NFTs collectibles have been spurned by CryptoPunks.

- **Sporting Moments**

Major sporting brands such as the NBA and ATP have come to be another type of NFTs. The rapid growth of this space is due to how those brands are now selling ownership of video clips from their key in-game events. For example, the clip of NBA star Zion Williamson of the player blocking a shot is one of the best-selling NFTs valued at $100,000. According to this theorized pattern, the values of such sporting moments will continue to rise over time.

NFTs and Cryptocurrencies

NFTs are generally built with the same kind of programing as cryptocurrencies. Both are represented as digital assets and are stored on top of the blockchain protocol. But that's where the similarity ends.

Nevertheless, the majority of crypto assets are virtual currencies, and like physical money, can be used as a medium of exchange. They're fungible in that they can be traded or exchanged for one another. The values of such currencies are equal; one USD is equal to one USD any time anywhere and one BTC is always equal to one BTC. This fungibility of cryptocurrencies makes them a trusted and reliable means of conducting transactions and deals on the blockchain.

That is not so with NFTs. Each non-fungible token comes with a digital signature that makes it impossible for an NFT to be equal to or exchangeable for another (here comes the idea of non-fungible). For instance, one

NBA Top Shot clip is not always equal to another simply because they're both NFTs.

Are you wondering what really gives value to NFTs? Please read on.

CHAPTER 2: WHY NFTS HAVE VALUE

Have you ever wondered why anyone would pay millions of dollars for a virtual plot of land in a virtual estate? What about *Axie Infinity,* a game that incorporates aspects of cryptocurrencies as well as NFTs that have sold for up to $800,000? You would understand better if you know why NFTS have their value.

Why Collectibles Have Value

The value of an NFT depends chiefly on its type and those associated with it. Factors such as content, utility, and availability or scarcity also affect its value. Thus, the NFTs value is highly subjective.

A token is worth as much as someone is willing to part with to have it. In order words, these NFT collectibles have value simply because there are people willing to pay a lot of money to own them. The contributions of a famous artist or social media influencer can place a price of hundreds, thousands, or even millions of dollars on an NFT. A raw fact is that some artists are actually making millions by selling NFTs.

Another thing that adds value to collectibles is that they're fun and awesome to collect. Many are just passionate about collecting art and adding them to their collections. For others, buying NFTs is a way to connect with and support their favorite artists so that artists-fan connections can be more direct.

Speculators on assets also add value to collectibles. They're sure that they could make fortunes on a particular NFT as they hope it would someday be bought for more than the price they're willing to pay for them now.

Added to this is the status symbol an NFT gives to the owner. Famous people, including celebrities, often buy art as a way of showing their status. For instance,

Eminem, Post Malone, and Jimmy Fallon readily paid thousands of dollars to buy the *Bored Ape Yacht Club* (BAYC) avatars. The joy and satisfaction derived from such purchases are good reasons to go for them.

Other reasons include digital identity that is kept secure, intellectual property and patent, and project-specific utility.

The Problems With Traditional Art

There is a crossover issue between the traditional art marketplace and the NFT marketplace. NFTs came to the limelight when the art world was suffering from a kind of Gordian knot. It's been a bit challenging to establish a younger generation of art enthusiasts who would be willing to sustain the tradition of art when the current artsy generation is all gone. Web-based galleries, social media, and online auctions have been looked at with just a little success. The art world hasn't got its fair share of, or enjoyed deserved space in, the virtual world's exponential growth.

Fortunately, the NFT process which includes launching or minting, buying, selling, and trading has come to be a game-changer for art. While the traditional art world can present a tough and exhausting challenge for introverts, the virtual world is easily shaped to match collectors' desired level of communication, thanks to its anonymity. Thus, online galleries that focus on NFTs can conveniently bridge these two worlds and connect artists with the virtual world. Artists display their work to a global audience, while art fans and collectors have an innovative, playful, and comfortable way of buying art.

Of course, online galleries exist already. But they can't cope with the desires of modern users without the added mechanism of minting and NFT trading. With the addition of these, there now exists dynamics to address this modern challenge in the industry.

The Problems With Collectibles and Memorabilia

Collectibles are items worth far more than they were initially sold for because of their rarity or popularity or both. The NFTs memorabilia keep records of those collectibles worth remembering alive. It's easy to confuse memorabilia with collectibles while you'll be excused. However, they are two very different classifications which can further be confused with antiques. Collectibles are purchased just for fun, memorabilia serve a purpose, namely to remind you of a memorable event. Things like sporting event programs and souvenirs, including Olympics programs and World Series pennants are regarded as memorabilia.

But there are some problems with these, such as;

1. **Highly volatile values:** The risk of destruction of the assets you've brought at a great cost is higher.

2. **A lack of income generation:** When you purchase these collectibles for speculative purposes, you won't get any investment income or dividends until you're able to sell them.
3. **Lack of liquidity:** You may always hear stories about how someone spent a few dollars at a garage sale or found something in the attic and sold it for a fortune. This happens less often. You'll need to spend a lot of money on valuable items if you really want to make money in collectibles.
4. **High transaction fees and costs:** Though third parties have been sidelined in dealings in NFTs, the fees and costs associated with selling collectibles are substantial enough to be ignored.
5. **Prevalence of counterfeits:** The value of NFTs collectibles is in their uniqueness. Unfortunately, the risk of owning counterfeit memorabilia is quite high because, in the collectible

market, counterfeits are everywhere.

Digital Art Before NFTs

You're likely familiar with the old axiom, "Artists only get rich after they're dead." That was the situation digital artists inherited from traditional artists. Thanks to non-fungible tokens, this is changing as artists, content creators, musicians, and photographers have got a platform for digital protection and profit from their work.

Before the introduction of blockchain technology, digital artists were usually shut out of the traditional art market. It was so tough to convince collectors to cough out their hard-earned dollars for a .JPG or .MOV file that is stored on one thumb drive; many commercial galleries didn't even care to try the herculean task. In those days, before NFTs, digital art wasn't that profitable in the traditional art marketplace.

A lack of protection of intellectual property is another clog in the wheel of progress of digital artists. The ineffective way digital artists have used to safeguard their art was by slapping on a watermark to inscribe ownership. Unfortunately, these have been messed up by Photoshop and other photo editing tools which have made those relatively easy to remove. Before the introduction of blockchain technology which is changing the narratives, digital artists were being confronted by varied issues too numerous to itemize.

Thankfully, a variety of digital assets are nowadays being minted as NFTs. The market has subsequently experienced exponential growth in recent years such that if you venture into the market these days, you'll be among those who laugh last and best.

The True Advantages of NFTs

As you continue to learn about NFTs or trade them, you'd want to enjoy their true advantages. In addition to telling you why people run after NFTs, here are the biggest benefits of investing in collectibles:

1. **You can purchase items that excite you**

People usually get into specific collectibles as a hobby, apart from being passionate about making money. Thus, antiques, comic books, fine art, model trains, and sports memorabilia are good examples that have been bought because of the excitement they give the buyer.

2. **Proof of ownership**

Since they're on a blockchain network, NFTs ownership can be associated with just a single account. While the real offline assets are largely indivisible, tokenizing them as NFTs has made them available to a group of real owners. At the same time, that ownership advantage of NFTs protects buyers from the concerns of fake NFTs. You can't just download any image, say a Mona Lisa painting for instance, from the Internet and claim to be the owner. The NFT made that impossible by ensuring that you really own what you're having.

3. **Authenticity**

The records associated with NFTs are unique since they are created on a blockchain. This unique state and authenticity create values for these tokens. Creators may choose to issue only a specific number of these tokens, to create scarcity or create multiple replicas in case of tickets. Since blockchain-based NFTs are immutable, they are immune to modifications, replacement, or removal.

4. Transferability

NFTs are pretty easy to trade freely on particular markets with a wide range of trading options. The challenge of 'walled gardens' in the case of games, for example, has been resolved. In this case, game developers may decide to issue in-game items as NFTs. It's then possible for players to hold these in their digital wallets. Players could use those in-game items even outside the game or sell them for a profit.

5. Economic Opportunity

The fragmented nature of NFTs makes them feasible in the world of digital content. Thus, content creators no longer need to lose their royalties to other platforms. This benefit of NFTs could lead to the development and expansion of a completely new creator economy that focuses on content creators' profits. The ownership is integrated into the content only which makes the funds from the NFTs go directly go to them. Even if a new owner sells the NFT, by setting up a smart contract, the original creator could receive royalties since the NFTs metadata includes the creator's address.

6. **Diversification Benefit to an Investment Portfolio**

Just like traditional investments like stocks and bonds, NFTs have distinctive qualities and offer benefits to an investment portfolio beyond what has been seen. Adding collectibles as an asset class is an option for investors considering further diversification. And as you're now aware, NFTs collectibles have appreciation potential.

NFTs Aren't Perfect

How I wish we can continue to speak glowingly of NFTs without any challenges! But they're not perfect; not without challenges. Here or some of the downsides of NFTs:

1. **NFTs Are Illiquid and Volatile**

Given its relatively immature state and a lack of a lot of historical data to research, the market for NFTs isn't very liquid. As they're not yet widely understood, the number of potential buyers and creators is small. Trading them, especially during periods of distress, can be very difficult.

2. **NFTs Aren't Income Generators**

They don't generate as much income as rent-generating real estate, dividend-paying stocks, and interest-bearing bonds. So, they don't easily offer their owners income potential. The RoI in NFTs like antiques and other collectibles aren't something you should count on yet since they're based entirely on price appreciation.

3. **NFTs Can Be Used to Perpetuate Fraud**

While there's no doubt about the integrity of a blockchain, NFTs can be used to perpetrate and perpetuate fraud. Imagine the shock of artists who have found their work for sale as NFTs on online marketplaces without their consent. It's a matter of creating an electronic image of the original work, attaching a token to it, and putting it up for sale on a virtual marketplace. He can get away with it as there as there's no link to the original work.

CHAPTER 3: HISTORY OF NFTS

When Andy Warhol's NFTs sold for over $3 million around the second quarter of 2021, the world of digital art was shaken. The news drew unusual attention to the disruption being brought about by blockchain technology. However, that has already been dwarfed by Beeple's cyberpunk since the time Beeple sold an NFT for $69 million.

All these should arouse your interest in the history of NFTs and this is what we shall discuss in this chapter.

The Origin

By the way, how did this technological phenomenon come about, and from where? The modern history of NFTs and the pioneer creator date back to Kevin McCoy, who began it on May 3rd, 2014. He minted his token named Quantum way before the crypto art market exploded.

The idea now described as non-fungible tokens has been conceptualized (though without the term) long before the coinage of the term. It's a product of several experiments with Bitcoin and the blockchain itself. The desire to produce something rare or unique actually was of .bit domains. The project BitDNS was initiated when the "Alternative Bitcoin" called Namecoin was introduced with the goal that the coins would make for keeping a registry of .bit domains with the view that there was a possibility of keeping a unique record of data in Bitcoin.

The fast-moving world of NFTs has rapidly changed a lot. In just a few years, NFTs have metamorphosed from merely a concept of what can be achieved with blockchain technology to a billion-dollar industry capturing the imagination of tech and art enthusiasts worldwide.

Now, back to the McCoy non-fungible token; it's a pixelated image of an octagon filled with denoting arcs, circles, or other shapes that share the same center. The larger shapes surround smaller ones and hypnotically pulse in fluorescent hues. This one-of-a-kind "Quantum" art piece has been put on sale for over seven million dollars.

Andy Warhol Presents Pop Art

Before Andy Warhol's artworks started making sales as NFTs, he pioneered the transformation of commercial icons such as *Campbell's Soup Cans* into art. He has reproduced hundreds or thousands of pieces in the studio "The Factory." Thus, it's only natural and fitting for his works to be tokenized now that crypto is becoming a new medium for commercializing and trading art.

The American artist is widely recognized by his audiences and his critics alike as the leading figure in the American Pop art industry. Warhol's depictions of consumer goods and celebrities portray him as being fascinated by materialism and mass media and he fundamentally altered the art scene and its relationship to consumerism.

Down with the infliction of the neurological disorder called Sydenham's chorea, the bedridden Warhol occasionally entertained himself by experimenting with silk-screen printing in addition to movies and radio programs. His technique is now synonymous with his oeuvre having had his art skills honed by his study of commercial art at Carnegie Mellon University.

Before his sudden death at 58 as a result of complications of gallbladder surgery in February of 1987, Warhol's financial and critical successes, including the $105 million sales for *Silver Car Crash Double Disaster* in 1963 and the $81.9 million sales for *Triple Elvis* in the same year, were breathtaking. These consolidated his role as an American Pop art idol nationally and internationally.

Beeple's Cyberpunk and NFTs

Let's talk more about Beeple and his mammoth NFTs sales and his free cyberpunk visual effects. Beeple, whose real name is Mike Winkelmann, is a visual effects artist, designer, and animator from Charleston SC.

He generously creates a wide variety of stunning digital artwork, short films, and free-to-use abstract Creative Commons video loops. Often called VJ Loops, they're intended to be used as visual sequences to complement the musical performances of bands, DJs, and other electronic music artists.

Beeple's distinct style features flashy neon, high-tech, subtle film grain, abstract structures, reflective, and all the other trappings that are now being associated with the whole cyberpunk, blade runner aesthetic. Beeple is one of the earliest creators of the 'everyday' concept—a challenge or competition where artists are required to produce one work every single day for a long stretch of time.

Thus, Winkelmann has been able to keep his "everyday" going continuously for 10 years. This is unbelievably impressive in light of the complexity and quality that many of these works exhibit. But the beauty of this is that when you do the same thing for so long, so frequently and repeatedly, it's natural for it to become second nature.

While many artists are very passionate about protecting their methods, workflows, trade secrets, and intellectual properties, Beeple releases many of his works under Creative Commons. He also makes many of his project files downloadable for free to enable others to see just how he creates his 3D models, effects, animations, and more. With minimal restrictions, anyone can use his work. You are only required to make an attribution to the artists.

The Story of Digital Art

The story of digital art can be traced back to the 1960s when artists began experimenting with computers. Credit goes to John Whitney for the creation of the world's first computer-generated art. He's considered one of the "fathers of computer graphics," as he used mathematical functions to transform visuals while making his artworks.

In 1967, Engineers Billy Klüver; Fred Waldhauer and artist Robert Rauschenberg;

and Robert Whiteman formed a group named Experiments in Art and Technology (EAT) to promote collaboration between art and technology. The result was a series of installations and performances that experimented with innovative electronic systems, albeit not strictly digital. They nevertheless laid the foundation upon which a new type of art would be built when the technology is fully developed.

The groundwork laid by EAT was utilized by other conceptual artists to promote technologies. Allan Kaprow's creation of *Hello* in 1969 was an example of this. A group of people was able to interact by using T.V. monitors.

The actual first digital creation was made by Frieder Nake, an artist trained as a mathematician in Stuttgart, Germany. Nake entered an algorithm directly into an ER 56 computer that mathematically interpreted one 1929 Paul Klee painting. The room-size computer ran a flatbed drawing machine to produce many images which, according to

London's Victoria and Albert Museum, were the "most complex algorithmic work of its day."

And in 1970, artists began exploring the results of this connectivity, recording equipment, and nascent computers. When mobile phones and the internet were introduced, the idea of universal communicability was later discovered and thus started the era of development in technologies. The Apple II computer which allowed the execution of color graphics for the first time on the PC screen was introduced.

It was after a decade, in the early 1980s that the term "digital art" was first used following the creation of a paint program by computer engineers which became known as AARON. This robotic machine was designed to make a large drawing on sheets of paper placed on the floor.

In 1985, Warhol came to the limelight, creating digital art on an Amiga 1000 to help in advertising the computer. Thanks to artist

Cory Arcangel who worked with the Warhol Museum and the Carnegie Mellon University Computer Club, the 28 digital "experiments" were recovered after getting lost.

As the Internet and digital technology improved in the 1990s, digital art became more accessible to artists and viewers. Download to computers was made possible and digital artists were for the first time able to manipulate images they had filmed with a video camera.

The increase in Internet use since the millennium has made anything possible in the world of digital art.

This is what marketplaces are capitalizing on to facilitate the sales and purchases of NFTs as we shall see in the next chapter.

CHAPTER 4: NFTS MARKETPLACES

As the NFTs are gaining popularity in recent years, with the market crossing a $40 billion mark in 2021 and holding stronger in 2022, you may be looking to get involved in the NFTs market. In this chapter, we'll discuss the best NFT marketplaces so that you can choose a reputable and cost-effective, and best platform and NFT marketplace for your investment needs.

Let's carry along those who may not fully comprehend what NFTs marketplaces are. Simply stated, they are online platforms where NFTs are bought, sold, or minted. They can be likened to vast online stores where visitors can browse the selection of items listed for sale and make purchases. Remember that NFTs are hosted on the blockchain. Thus, the vast majority of these marketplaces require users to have a crypto wallet where post-purchase NFTs are stored.

The following are the top 10 NFTs marketplaces.

1. **OpenSea**

Opensea is a top NFT marketplace with a lot of asset selection. It's the largest in the world in terms of trading volume. Many people believe this platform is the best place to buy NFTs as the first to bring peer-to-peer NFT trading into the mainstream. Since its launch, it has reached over one million users with over $6.5 billion in trading volume as of the last check. This marketplace, built on the Ethereum blockchain, offers various NFT types, including art, domain names, and in-game items among others.

Signing up on OpenSea is as easy as connecting your MetaMask wallet. Of course, it supports other wallets such as Coinbase, Bitski, and Formatic. Buying an NFT or uploading your own creation is simple as well. To buy, simply browse through the various collections or search for something that takes your fancy and put in an offer. Just wait for it to be accepted. And to sell, upload your NFT creation and hit the "Create" tab. Next, connect your wallet as a creator. Then upload your NFT and fill out the description. You'll learn more about these later.

2. Rarible

This is one of the best and leading NFT art marketplaces for investors. The Rarible platform hosts a wide range of digital art selections, along with domains, games, music, photography, and even Metaverse items. The charge is a 2.5% fee to both buyers and sellers. Though this puts it at the higher end of the cost spectrum, it has multi-chain support and can host NFTs built on Ethereum, Flow, and Tezos.

One great thing about Rarible is that you can buy an NFT there using your credit or debit card while it supports over 190 countries. This one of the biggest NFT marketplaces also supports several wallets, such as Coinbase and MetaMask. Thus, you can store your assets safely post-purchase.

3. Nifty Gateway

This is the best NFT platform for rare drops. It's one of the first waves of big NFT marketplaces. Brokering one of the first multimillion-dollar NFTs sales, Beeple's *Crossroad* for $6.6 million, Nifty Gateway made headlines in February 2021. Three months later, this platform had hit $300 million in merchandise value.

It plays host to artists such as The Weeknd, Grimes, and Eminem. Offering a heavily curated selection of NFT drops, it's a marketplace where you can browse curated tokens, and verified and unverified artists. The process of getting started is easy. Simply register on the site, and if you're a seller, you'll need to verify your identity.

4. SuperRare

SuperRare is another NFTs marketplace that is more heavily curated. Billing itself as a social platform aiming to encourage the creation and collection of crypto art, it positions itself as a higher-end alternative to Rarible. While Rarible's interface throws on you a dizzying array of flickering GIFs and live auctions, SuperRare has a more pared-back, minimalist appearance altogether. Working closely with artists, it requires that work be submitted and approved before it can be listed, thus placing quality over quantity.

To sign up to SuperRare, you'll need a supported MetaMask or Formatic and a few other wallets it supports. You also need to create a username and password linked to your wallet address, and this is done within a few seconds.

5. Wax (AtomicHub)

Wax is a widely-used NFT marketplace launched in 2017 as a blockchain designed to improve the efficiency of e-commerce transactions. It's compatible with EOS, leveraging blockchain power to compete with Ethereum. The WAX NFT marketplace, also known as AtomicHub, is called SimpleMarket. It offers more than 140,000 digital assets.

The vast majority of NFTs on WAX are denominated in WAX as the network's native token. But there's also a popular secondary market for WAX NFTs on AtomicHub. This has no gas fees. Atari and Sony are among the high-profile brands that have already used the WAX blockchain to mint their NFTs and this adds more credibility to the marketplace.

6. Foundation

Foundation is a premium NFTs marketplace for exclusive art with a total trading volume exceeding over $79 million. It has proved to be a creative playground for artists, playing host to a number of notable sales, such as the NFT of iconic meme Nyan Cat and Edward Snowden's first NFT. You'll also find there an audiovisual digital collectible produced by Richard D. James, better known as Aphex Twin.

You simply need a connection to your MetaMask or other software wallet using WalletConnect to operate Foundation's marketplace and start placing bids on timed auctions as if you're on a regular auction site.

7. NBA Top Shirt

NBA Top Shirt is great for sports NFTs and it's among the first series to become popular among the wider public. It features a set of digital trading cards having NBA video highlight clips that are stored in your secure encrypted wallet after purchase. In this blockchain-verified wallet, you can view or resell them on the NBA Top Shot Marketplace.

The creators Dapper Labs make getting started simple; you just need to link your Google account to Dapper. Then, enter a phone number for SMS authentication. You thereafter have a few account setup steps, including selecting a favorite team, to go through. You can pay with an existing crypto wallet, your Dapper balance, a Flow wallet, or a credit card.

8. VeVe

VeVe is a public-oriented mass NFTs marketplace. It was created by ECOMI, a

Singapore-based cryptocurrency company. The ECOMI is based on the cryptocurrency ecosystem known as the OMI token. It has a user base numbering in the millions. You don't need a deep understanding of crypto, wallets, or the underlying blockchain technology before you can create a mass-market product.

9. KnownOrigin

KnownOrigin is one of those NFTs marketplaces desirous of providing a more curated, gallery-like experience for the discerning NFTs connoisseurs. It has all of its artwork files held on IPFS. This gives a measure of protection (beyond the rest) for the underlying assets. This marketplace is firmly focused on digital art, shunning the crazier aspects of the NFTs world. Thus, you can't find wacky avatars or cute monsters on KnownOrigin.

KnownOrigin is easy to use. It supports wallets like MetaMask or Formatic.

10. Myth Market

Then comes the Myth Market, a series of convenient online marketplaces that support various digital trading card brands. It currently features GoPepe.Market where you can buy GoPepe trading cards, GPK.Market for digital Garbage Pail Kids cards, Heroes.Market where you'll see Blockchain Heroes trading cards, KOGS.Market to get for KOGS trading cards, and Shatner.Market where you can acquire William Shatner memorabilia.

This marketplace is built on the WAX (Worldwide Asset eXchange) Ecosystem. When it comes to video games and entertainment networks, Myth Market is vying for the leadership slot in the party of NFTs marketplaces. Well, it has boldly claimed to be the "safest and most convenient way to create, buy, sell and trade" non-fungible tokens to anyone anywhere in the world.

It would be challenging to place one NFTs marketplace above the other. The fact is that one that is doing so great in an aspect is lacking in another aspect and area, and vice versa.

CHAPTER 5: CREATING AND MINTING NFTS

Up till now, you have seen several successful NFT projects. This might cause you to wonder how NFTs are created and minted. I understand that despite the explosion of NFTs popularity, not many people understand how to mint an NFT. So, let's devote some of this piece to that. Yet, creating and selling NFTs, as you've seen, is an exciting and new way for digital artists to mint ownership of their works and get them bought at an NFTs marketplace.

But then, creating and selling NFTs requires the use of complex blockchain technology. Don't panic; you don't have to be a wizard to create your own. It's the easiest thing to do. Denise Schaefer, co-founder of Surge, a blockchain education platform, says: "There's a big misconception that you have to be technical in order to participate in crypto. . . . But I look at NFTs as a fun entryway into the space that doesn't require coding skills when minting in marketplaces like OpenSea."

Aspects of Your Own NFT

You start by deciding on your concept. Know what you want to tokenize and how you want it to look. Then, come up with a befitting name. This requires that you have good art, video, music, or other valuable to sell.

The next thing is to determine the NFT platform where you will mint your NFT. There are tons of such platforms around. They include the NFT marketplaces mentioned in the preceding chapter of this book. Many more continue to spring up with the continued growth of interest in this type of asset. That gives you a lot of choices.

While you need to research to determine the platform that fits your needs, I want to recommend Opensea. This is because of its popularity and ease of use for all types of NFTs.

Cryptocurrency Wallet Needed

Since you need cryptocurrency to trade on NFTs, at least to be paid when one is bought, you must create a cryptocurrency wallet. The crypto wallet does for cryptocurrency what a bank does for fiat currency. So it's a lot like a digital bank account. It has information about how much of each crypto asset you own. It comes with your unique identifier called a key.

Examples of such wallets used for NFTs are Metamask, Nano X, Trezor One, and Exodus among others. MetaMask is a Google Chrome extension and is one of the most used wallets, especially for beginners in NFTs. It's stable, reliable, and acceptable across most NFT-used platforms. However, it doesn't have the same level of security as the other three mentioned earlier.

Creating an Opensea Account

Let me begin by telling you that the blockchains used by Opensea are Ethereum and Polygon. Granted, the Ethereum blockchain is known for charging high gas fees. OpenSea now has a more convenient mint option that allows the creator to upload his NFT, mint it to his profile, and list it for sale free of gas fees. This cost would be borne by the collector when he buys it. The creator fee is 2.5% of your sale.

On the OpenSea account, you need an ETH wallet such as MetaMask, Coinbase, or several others. Here are the steps:

Browse opensea.io and go to your profile icon at the top right corner. Around the same location, click the "Create" icon. That'll take you to the NFT item creation page where you'll be allowed to upload your NFT file with the appropriate name and description. There are specific acceptable names, descriptions, file sizes, and formats.

Next, you'll be prompted to customize your NFT further. Doing this, you can place it on an existing collection, or add properties, stats, levels, and even unlockable content.

Here, you can select the blockchain you intend to mint your NFT on.

Once you're through with customizing your NFT, click the "Create" tab at the bottom left corner. The next thing is to start minting and selling.

Summarizing Minting an NFT

Let's illustrate all that using OpenSea, taking it step-by-step:

1. **Connect Your Wallet to OpenSea**

Go to opensea.io to see the marketplace and connect as described above. Remember most successful NFTs are part of some collections. So, create a collection instead of a standalone NFT. Don't forget also that you will have to add all the details about your collection.

2. **Create an NFT collection**

To create your NFT collection, add

- Logo, featured, and banner images.
- Name
- URL
- Description
- Category (art, collectibles, music, photography, sports, trading cards, video, utility)

- Links (Discord, Instagram, Medium, Telegram, or website)
- Royalties percentage
- Blockchain to mint the NFTs (Ethereum or Polygon)
- Payment tokens—ETH/WETH on Ethereum & ETH on Polygon
- Display theme on OpenSea

Once you supply all those details, click on "Create" to create your NFT collection.

3. **Add items to the NFTs on OpenSea platfrom**

You do this by clicking on "Add item" on the collection page (top-right). Next, go to the "Create New Item" page to specify all the details of the NFT you're adding items to. The details include:

- Image, Audio, Video, or 3D Model
- Name
- Description
- External Link
- Collection
- Supply
- Blockchain

- Freeze metadata

Following that, just click “Create” to create the NFF.

For your NFT to be found on the explore page, you have to list it for sale. That’s will be discussed in the next chapter.

CHAPTER 6: SELLING NFTS

All the efforts you've put into creating and minting your NFTs will amount to nothing if you haven't successfully put them up for sale. You may be one of the NFT creators looking to sell NFTs as they're creating functional tokens serving full-on utility services. One of the earliest steps to take in selling your NFTs is to determine the ideal marketplace for you, and you already know those marketplaces. Let's go through the steps now:

1. **Create your exchange account**

Don't forget that what you're dealing in are blockchain-based assets. You need a wallet in one exchange to create, send and receive them. Your wallet must be funded to enable you pay fees associated with your transactions. As you know, most NFT platforms are built on Ethereum. That's why you need an Ethereum-wallet funded with ETH, their native currency.

The account has to be on a centralized exchange. The standard procedure is that you go through the KYC verification process which requires that you provide your ID information for the exchange to ascertain the identity of who's trading on their platform.

The big question then is, which exchange should you be using? Yes, there are plenty of options to pick from, each with its pros and cons. However, most experts recommend the Metamask wallet which is an extension on Chrome and other browsers. Metamask allows you to connect with a marketplace and use related services directly from your browser.

Please note that while installing and running Metamask for the first time, you'll receive a 12-word recovery phrase. Make sure these are saved in a secure offline location. Then, proceed to copy your ETH address.

2. **Buying cryptocurrency and funding your Metamask wallet**

Buy your Ether (ETH) with which you will fund your wallet. This means you buy from a cryptocurrency exchange or broker that supports your Metamask ETH wallet address and send it there. If you deal with some brokers, you only have to provide your wallet address on the purchase and they just transfer the coins to your address. However, if you use an exchange, the ETH will have to be withdrawn after the purchase before you fund your Metamask ETH wallet with them.

On some platforms, you will be able to complete the purchase using your credit/debit card or bank details.

3. **Determine the price of your asset**

Remember the maxim, "NFTs are worth what people are willing to pay for them." Nevertheless, certain factors influence their value and pricing. These include:

30-day average price: This refers to the historical price trend of an NFT. Your target price should normally be close to the average value of similar NFTs over the past 30 days.

Rarity: Certain NFTs are rare and this drives up their prices. An example is the Bored Ape Yacht Club NFT collection. If your NFT also is rare, you can price it above the historical average.

Utility: If your token can be traded for a virtual item, or if buyers can be a part of an exclusive club that will receive extra perks, such as discounts on future purchases, the utility of your NFTs will be higher, and you can place a higher price on them.

4. **Get your artwork ready**

Of course, by now, you likely have your artwork prepared. You can browse through the marketplaces discussed in the preceding chapter to get inspiration on how you can prepare a unique top-selling art, video, audio, image, or 3D model.

5. **Selling your NFTs**

You're good to go selling the artwork. Let me use OpenSea which I recommended earlier to illustrate how to go about selling. Follow the step mentioned in Chapter 5 about creating an OpenSea account and minting.

Thereafter, click on the “Sell” tab on the top right and you’ll be taken to the list page.

Here is where you will choose your price and type of sale.

You have the option of Fixed Price sale where you will establish the NFT price or Timed Auction Sales.

You’re also allowed to set a duration for your sale. The default options on OpenSea are 1 day, 3 days, or 1 week. However, you can set a custom duration using the calendar.

As mentioned earlier, you can fill the address of a specific buyer into the “Reserve for a Specific Buyer” field to reserve it for just that buyer.

Bear in mind the potential fees from the sale as listed at the bottom. They usually are OpenSea’s Service Fee and Creator Earnings.

To complete your sale, you’ll be asked to confirm by signing a transaction.

If you're new to OpenSea, you'll need to initialize your wallet first. Additionally, if the item you're listing was not minted on OpenSea, maybe it's through a custom contract, you may need an additional approval and signature to authorize OpenSea to trade your NFT on your behalf.

6. **Marketing your NFTs**

The job is not done until you have promoted your content. Fortunately, OpenSea and other marketplaces use social proof among other criteria to approve content creators. It's compulsory for artists to specify the number of social media channels they actively manage and share their NFT collections so that the public can see them, at least, once.

To properly market your NFT, you need to set a goal of letting your audience know what you're venturing into and let the word go out as soon as your first collection is minted. You may also release free edition NFTs to build your followership and sales record on the platform.

To be sure, minting and selling your first NFT may not be particularly smooth. Nevertheless, the effort and stress may well be worth it by the time your goal is realized. But how about buying NFTs? Read on to see.

CHAPTER 7: BUYING NFTS

NFTs are certainly the hottest commodity in the digital assets world right now. Thus, buying an NFT is an exciting prospect! If you've been keenly following this book to this point, it should be pretty easy for you to buy NFTs. Some buyers consider the extreme highs and lows of NFTs values a high-risk asset and are being scared off.

I can assure you, however, that buying NFTs is as easy as or even easier than selling. However, let's start from here: which NFT is the right for you?

Choosing the Right NFT

The NFT investor Mike Jelinek warns that NFTs have a high potential for rewards as well as risks, adding "there are buys you look at speculatively, that you want to flip, and ones that you want to hold onto. Understanding your budget and what you're willing to lose is key." His recommendation?

Set "short-term, medium-term and long-term goals for your investments."

Since the NFT market is relatively young and highly speculative, no one can beat his chest about an NFT's long-term value, for now. That's why Jelinek's advice is that your budget, your goal, and your situation should be the main determinant of the right NFT for you. As you reflect on your goal, consider the following types of NFT in addition to what we've discussed in the first chapter.

- **Single edition:** This is also known as a 1/1 NFT or 1x1 NFT. In this case, only one NFT is created by the creator.
- **Multi-edition:** In this case, the artist creates a collection or series of related NFTs.
- **NFT-for-resale on an active secondary market:** This refers to a situation in which you bought an NFT to be resold.

- **Utility NFT:** These NFTs tokenize physical items, such as real estate or an event ticket.

The Process of Buying

Once you determine the NFT you want and why you want it, you'll be able to determine the marketplace you can meet the seller. For the discussion about NFTs marketplaces, please read again Chapter 4 of this book. And if you want to remember the reasons for buying an NFT and their true value, you can go back to Chapter 2.

Do your due diligence when looking to buy an NFT to ensure you're buying from a reputable source. As you embark on your research, make sure to read reviews, compare prices, and check other conditions before making a purchase. Once you're sure of this, initiate the process by accessing the marketplace.

The process of buying an NFT starts almost the same way as selling namely;

- Register or create an account on the marketplace or exchange (if you're not sure about this, please refer to Chapter 6)
- Follow through the payment steps to buy the NFT from your account.
- Be patient as you wait for your transaction to process.
- Confirm the NFT's destination address in your account.
- Check to be sure that the NFT has been duly received by you.
- Install the appropriate application for your platform to receive the NFT.

A word of caution:

Popular NFTs can sell very quickly. Make sure your wallet is funded and remains connected before the NFTs drop. Note please, that buying an NFT doesn't mean you already bought the copyright. Unless that's part of the agreement between you and the creator/seller. Try to understand restrictions placed by the marketplace on the NFT you purchased.

Storing and Viewing Your NFT After Purchase

By now, I want to assume you understand that you'll need a digital wallet in which to store your newly-purchased asset. Please, take care to choose the digital wallet compatible with the NFTs you just purchased. The process of creating an account to deposit it into your wallet has been discussed earlier in the book.

As the proud owner of an expensive digital asset, that can be traded, sold, or used in some way, you need to periodically view your NFT. That's also simple; just know the blockchain your NFT is stored on. The seller or the NFT's website can tell you if you don't know or forget. This may attract a fee in some cases.

Caveat Emptor

You should note that you're responsible for your security when buying here. If you're not new to trading online, please engage all you have learned about keeping your information and cryptocurrency safe when buying NFTs. This includes using two-factor authentication; starting the transaction with a small amount of money to be sure everything is working. There are scammers everywhere and are almost difficult to spot and beat.

However, there are a few signs to look for. It's important to check the seller's reputation. A seller with a lot of negative reviews, and little or no feedback, may not be very safe to deal with. Another thing is the NFT itself. Take a close look at the quality image; if it's low or seems otherwise unprofessional, beware. Finally, check the price; it's too cheap, it's likely a trap to lure you into a scam.

Watch out for the following sharp practices in this market:

- **Phishing scams**: These fake links and pop-ups advertise things like new NFT projects and drops on social media.
- **Catfishing:** These are fake marketplace websites or social

media accounts. Celebrities can be impersonated as advertising NFT drops and collections.

- **Pump-and-dump schemes:** These are hypes built around an NFT collection to pump up the price. After the top-tier investors have quickly cashed out, the bottom-tier investors are left with worthless assets.
- **Counterfeit NFTs:** These are people's work being sold as their own original work.

Besides these cautions, buying NFTs is an interesting endeavor. But there are legal aspects of NFTS, as we shall see in the next chapter.

CHAPTER 8: LEGAL ASPECTS OF NFTS

The creation, buying, distribution, and ownership of NFTs are new phenomena. These raise a plethora of legal issues that have not been captured in the laws of many land and pronouncements. In the NFTs field, there are ambiguities in the application of extant laws to copyrights, data protection, financial liabilities, contract, criminal aspects, and many more.

Are They Securities?

For instance, are NFTs securities? The US Securities and Exchange Commission classifies non-fungible tokens as securities in most cases. But if you consider the fact that a security is generally a claim on the future proceeds from another person's works, you might agree completely with that classification. After all, NFTs are usually the products of works that have already been undertaken. So most NFTs around can't readily pass for securities.

Nevertheless, NFTs can become subject to Securities and Exchange Commission oversight straightforwardly through fractionalization. When multiple investors buy portions of an NFT instead of one person having to own the token, it can be viewed as security. Even then, the argument still continues on both sides. While SEC is embarking on a crackdown on 'unregistered exchanges' dealing in NFTs, the exchanges are fighting to continue trading independently of the commission.

Intellectual Property Rights

Anyone buying NFT isn't buying the digital work itself. What buyers are buying is a collection of code known as metadata. This links to the 'true' version of that work which gives the issuers the ability to control tightly purchasers' use of any Intellectual Property Rights associated with an NFT.

It might first seem that once a claim to a unique piece of content has been sold, copyright has been assigned to the buyer. However, copyright and other IPR typically are retained by the issuer while the buyer is granted the right to display the underlying asset. The legal imbroglio arises from how and whether IPR is licensed through the sale and transfer of the NFT. The issuer will want to assure that his valuable brand is protected such that effective remedies have been stipulated in case of any misuse of his IPR.

As a purchaser of an NFT, what you will own depends on any embedded coding or smart contract in the NFT, or the terms of sale, as it applies in traditional contracts. For instance, NFT creators may program an NFT to generate an automated ongoing payment of royalties on any resale of the tokens. It may be possible for the issuer to track resale since they'll be logged on the same blockchain.

Each marketplace and even each individual creator may have different terms which buyers must always check thoroughly prior to a sale to know what exactly they're purchasing. The metadata written into the blockchain contains information about the location of the original work and the owner of that particular version. Even if anyone else downloads and views the digital artwork, nothing changes.

The complexity arises when someone creates and sells an NFT of existing art, video, music, or others' intellectual property in which they have no rights of ownership.

Taxes

Another aspect of NFTs where the law is struggling to regulate its increase in popularity is taxation. Globally, there's still a dearth of legislation and guidance as to how to make it a taxable trade or ensure the taxability of proceeds of NFT transactions. While governments globally are closing in on regulating cryptocurrencies, NFTs fall under a slightly different category of digital assets. Since they are separately identifiable, they can't be pooled for Capital Gains Tax (CGT) purposes.

While you can say NFTs are taxable, it's particularly difficult to determine where NFTs are situated for tax purposes. Yet, the same laws that apply to fungible crypto can apply also to non-fungible assets. Investors in them have their revenues accrued from sales taxed as property which is subject to the capital profit tax.

For instance, if you buy an asset with fungible crypto, it's a disposal of the crypto and it incurs a capital gain/loss. So if you purchase a collectible card at Opensea using ETH, you'd incur a capital gain and be forced to pay taxes. And a seller offering an NFT for crypto would incur a capital gain/loss.

Trading a piece of art for another also attracts some taxes.

Many more questions and situations are still left regarding NFTs and taxation. It becomes more complex as taxes and related regulations vary from one country to another.

Right of Privacy

There is the issue of anonymity in the sale of NFTs which is unlikely to involve the disclosure or exchange of significant personal data. But the security of even the little data and NFT transactions can become paramount. A blockchain may prevent you from exercising fully your data protection right, the right to refuse to disclose some of your personal data or delete/erase what you already supplied.

NFTs that include private data might violate certain principles and the spirit of some of the data protection laws.

The seemingly unrelenting rise of NFTs over the last few months is indicative of the fact that it will take a while before the law can fully catch up with it. The legal ambiguity surrounding different aspects of trading in digital artworks via NFTs is not limited to the examples considered above. We can also analyze other areas such as data protection, contracts, estates and succession planning, questions about money laundering, and data hosting and information safekeeping, among others. It will take some time before legal disputes about NFTs and digital artworks will be decided in the courts.

This begs the question of what is the future of NFTS. That will be the focus of the next chapter.

CHAPTER 9: THE FUTURE OF NFTS

The future of non-fungible tokens is becoming clear as they're now being used by various industries. People now spend billions of dollars in transactions and a strong community now supports the concept. Some experts nevertheless still doubt that creating, collecting, and trading NFTs for profit is a worthwhile investment. Since it's impossible to curtail the acceptability, they concede that it's better to acquire NFTs just to fulfill personal desires, not because you want to join the fray.

The Metaverse

NFTs are now a tool of communication business in the Metaverse. Theoretically speaking, an NFT is usable throughout the Metaverse, virtual worlds where happenings in the real life are represented on the internet.

Many companies are relentlessly working on creating virtual reality Metaverses. There's an endless possibility within the Metaverse worlds. There, you can design your life, design your avatar, interact, work, and play with real people in virtual communities, and explore new worlds using virtual reality headsets, smartphone apps, augmented reality glasses, or other devices. The movie *Ready Player One* is an example of what the Metaverse is.

As Metaverses become where you and your family can visit, NFTs will become an essential part of Metaverses. It'll serve as the building blocks for assets utilizable across all these worlds. The walls of your virtual homes will bear original NFT art and prints of duplicated digital art collectors sold for you. You can invite friends over to your immersive digital home to come and enjoy your NFT music and see your NFT art.

In the future, NFTs will enable the Metaverse to also blend with the physical world. We might see physical venues with digital counterparts in virtual worlds or NFT art galleries that exist both in the physical world on digital displays. Music concerts might be held both in a concert venue and a virtual bar while tickets will be sold as NFTs.

Big Tech Placing Big Bets on the Metaverse

Big tech companies are embracing this potential of Metaverse. When all described above fully materialize, we can potentially interact with each other in 3D environments, not only for workplaces, social connections, and gaming, but also for exercise, education, commerce, health, and more.

The earliest divers into this future are:

Decentraland: A digital land.

OnCyber: 3D NFT Galleries.

Sandbox: Opening gaming Metaverse like Minecraft powered by NFTs.

VCCESS: Launching the first phase of their Metaverse as a social marketplace.

I don't need to tell you what Facebook is doing in the Metaverse. Twitter's recent effort at previewing how NFTs could be incorporated into its platform is symbolic. The success of this will enable users to connect their profile to a MetaMask wallet and their collectibles will appear on Twitter.

The Future of NFTs on E-commerce Platforms

Ordinary people no longer find it difficult to buy NFTs, thanks to their user-friendly features. The e-commerce platform is now growing with the increase in NFTs. The recent world events have shifted the attention of the world to these online sites where they can buy critical items. Thus, many e-commerce platform owners are considering how NFTs could be included to boost their value.

Many third parties are in for disruption, no thanks to NFTs being brought to the platforms that they're already involved in e-commerce. Bringing on NFTs would soon eliminate them. By merging NFTs and e-commerce, businesses affiliated with those e-commerce platforms will enjoy a high-level profit value against their items.

Consider a downside of conventional shopping that will be taken care of by NFTs. Yes, e-commerce platforms spare you the long-distance drive to pick up a single item or a long wait in queues. Yet, the goods don't get to you immediately. However, using NFTs can help you work around this. The digital version of the product can be given to buyers right away as a token of what is to be delivered soon. This can be the greatest of all gratifications from an e-commerce purchase.

The Future Is Bright

Going by what is currently playing out in the world of non-bankable assets, NFTs have come to stay. It's already demonstrating some potential in the following areas:

- **Membership**

NFTs will be used as digital membership tokens providing access to exclusive groups.

Consider what is being done with the Bored Ape Yacht Club (BAYC). BAYC NFTs serve as badges or tickets allowing holders to join a private Discord channel so as to access special members-only perks. The success of BAYC as a membership badge hinges on the exclusivity FOMO. Only time will tell if NFTs wouldn't be the standard entry protocol for groups planning on growing with input from their members.

- **Gaming**

From what we can see, the potential of NFTs in the world of gaming can only be anticipated as we're seeing more and more games being built around NFTs. Consider the heavy investment in digital avatars, statistics, identities, and statuses with more desire to spend considerable sums in upgrading online persona or unlock exclusive content that will allow them to play in new worlds. What about the play-to-earn model which is another way NFTs are changing the dynamics of the gaming industry?

- **Asset Management**

If NFTs continue to be used for digital art asset management as they're already being used, artists and performers may accept them as the gold standard to manage their rights and licensing. Now that collectability is almost non-existent following the proliferation of internet file-sharing, NFTs are now adding value to digital assets again through irrefutable proof of ownership. A musician, for instance, can create an NFT which contains nine musical tracks, making each an individual NFT. The asset management potential of this new type of NFT has no bounds.

- **Building Society**

It has been observed that NFTs offer a social stake in a development community. Purchasing an NFT has been a way for people to support the projects and groups that they believe in. Thanks to real democratization, contributors have a confirmed and established stake and thereby can become vote-holding members of DAOs

(Decentralized Autonomous Organizations. Furthermore, there's the possibility of users farming NFTs to be donated to certain non-project projects to raise funds.

Yes, only time will tell what will happen to NFTs in the future.

CONCLUSION

The situation of NFTs doesn't always follow this logic. The more they hit headlines the more confusing they get and the harder it becomes to understand them. Indeed NFTs can be so tricky to comprehend or deal in. Thus, I certainly understand you, and will definitely not count it a minus for you if you're still not clear about what really are NFTs and what they can be used for in spite of how much or little you've heard about them.

However, this book has done justice to it by giving you areas to make more research on as you delve into the world of non-fungible tokens.

Made in the USA
Monee, IL
05 November 2026

67380860R00059